The TXT Book

Your Guide To Social Networking

The Socially Connected

authorHOUSE®

AuthorHouse™
1663 Liberty Drive
Bloomington, IN 47403
www.authorhouse.com
Phone: 1-800-839-8640

First published by AuthorHouse   3/24/2011

ISBN: 978-1-4567-3476-3 (e)
ISBN: 978-1-4567-3477-0 (sc)

Library of Congress Control Number: 2011903562

Printed in the United States of America

Any people depicted in stock imagery provided by Thinkstock are models, and such images are being used for illustrative purposes only. Certain stock imagery © Thinkstock.

This book is printed on acid-free paper.

Special Thanks to

Jim Reme, Monmouth University Photographer for the photos and Bea Rogers, Monmouth University Assistant Vice-President for Academic Foundations/General Education for her editing, and

Alyssa Vasquez for her drawings at the beginning of the chapters.

Contents

Introduction

This book has been a team effort that began in a classroom at Monmouth University in West Long Branch, New Jersey. In the Fall of 2010, there were 24 students in this class entitled Social Connectedness in the Age of Technology. Students registered for this required First Year Seminar class based upon their availability and/or interest in the topic. The goal of the class was to talk about the way technology helps and hinders the communication that occurs between parents and children, friends, boyfriends and girlfriends, and all others whom we come in contact with during the day.

Within the course, we began to talk about social networking and what we felt were the "unwritten rules" of cell phones, texting and Facebook. We realized that there was a gap in information about social networking, and that it would be helpful if there was a book to guide social networking users on the expectations and basic questions. We realized we were all dealing with questions such as: when do you text and when do you call (especially between digital natives and digital immigrants)? When should you call a friend and when can it be a text? Have you wondered what it meant to receive a one word text answer, or worse yet, no response from a text? And what about changing your status on Facebook? How often, and what is the importance of doing so?

This book was a collaborative effort which used the course materials, electronic discussion boards and student insights to come up with a book that we hope is useful and enjoyable. This book is intended to be helpful and a fun take on the technology we have come to depend on. We acknowledge that there are other social media sites like Twitter, Myspace and LinkedIn, to name a few. We specifically chose to focus on the phone, texting and Facebook because they are so important to the members of this class at this time period. Please enjoy the information the way it is presented. Also understand that technology moves so fast the rules may change just as fast.

The students and professor for Social Connectedness in the Age of Technology, Fall 2010, Monmouth University (listed on the back cover).

Chapter 1

The Use Of Cell Phones

Believe it or not, cell phone usage has its own unique set of principles that apply to everyday use. This is because there are so many things a phone can do these days that the last thing they are considered is actually a phone! With all of this change, it can become rather difficult to keep up with what is and isn't acceptable in cell phone usage these days. Thankfully, we are here to present to you with everything you need to know about using a phone that should help you remain in the twenty-first century. Whether it is about texting or making a call, we've got your back!

Calling vs. Texting

There are many different situations when it may be inappropriate to talk on the phone and texting just may be the better option, but there are also times when texting could be the better choice, such as when you want to ask a quick question or have a quiet conversation. There is a proper time and place to send a text message and a proper time to make a phone call. Let's take a look at them.

When it's appropriate to call

□ **If it's something important**
If there is something very important or urgent that may take a long time to explain to another person such as grave news, personal matters, or congratulatory circumstances.

□ **If it's something that requires a lengthy explanation**
When something crazy happens to you and you want to tell your best friend right away, and it is going to be a long story, then a phone call is easier so you do not have to waste the time to send a long text message.

☐ **If you are in a hurry**
Calling is a good thing to do if you are in a rush and you only want to ask the person a quick question. With texting, they may take a few minutes to respond, which is not ideal if you are looking for a fast response.

☐ **If you want to have a full conversation**

- If you have plenty of time to have a long conversation (more than 45 minutes) with the person, then it is a much better idea to pick up your phone and call.

- **If the conversation is emotional**
A phone call makes it easier to hear emotional reactions and tell what the other person is thinking. With texting, it is much easier for the person to hide their true feelings about the topic.

- **If the person you want to communicate with is technologically phobic or completely clueless how to text**
You may need or want to talk to someone who cannot text fast enough to give you the information you want, such as a grandparent or your own parent. When that occurs, you may have no choice but to show respect and dial the phone rather than text.

When it is appropriate to text

☐ **If it is convenient**
Texting the person that you are trying to get in touch with can be much easier for many different situations: for example, if you are in a loud or crowded place and can't hear the other person on the phone.

☐ **If you are thinking about a special someone and want to send them a quick "catch-up" hello**
It is more appropriate to text the person if you do not really have the time to talk and have a long conversation, but you

still want to hear what has been going on in their life but you still want him/her to know you are thinking of them.

☐ **If you are busy**
Texting is a great thing if you are super busy and do not have the time to sit down and stop everything else you are doing. With texting, you can answer the person at your own convenience, and still be doing other things while you are talking to them.

☐ **If you have to be quiet**
The library is not the place where people want to hear you talking to someone on the phone while they are trying to study. Texting is silent and is a great thing to use if you are in a quiet place where it is not appropriate to talk on the phone. But don't worry about the sound of clicking keys; it all becomes white noise.

☐ **If you need privacy**
If you are in a room with a lot of people, texting is the easiest thing to do if you do not want everyone in the room to overhear everything you are saying.

☐ **If you are nervous**
Some people are too nervous to say exactly what is on their mind on the phone or in person. They may have an easier time saying some things that would otherwise make them feel uncomfortable to say out loud through texting.

☐ **If you are under pressure**
Texting gives the receiver the chance to think about what they want to say next, without feeling the pressure of needing to respond right away.

☐ **If you are socially awkward**
Many people prefer to just text the other person because then they do not have to deal with uncomfortable greetings or strange silences. It is much easier to get straight to the point with a text message. This is only a temporary solution, though. If you are socially awkward you will want to

address that in the very near future because you can't hide behind texting forever!

Simple Rules of Phone Etiquette

That guy; we all know him. The one who is sharing his dinner plans with everyone on the subway, texting or checking his e-mail at the movies, or fighting with his mother or significant other in a waiting room. In order to avoid being him, it is important to know and practice proper cell phone etiquette. By following the rules, you can be considerate instead of being a public nuisance.

Rule #1: Use an **inside voice** when speaking to someone on the phone; believe it or not they can hear you so there is no need to yell into the phone. This is rude and disruptive for the people around you.

Rule #2: Use careful **discretion** when talking about personal aspects of your life in a public place. People in the local coffee shop do not want to hear about how you and your significant other are fighting or how drunk you got this weekend. Also, be mindful that while you have stopped to talk or text, everyone around you has not. Fielding a call while shopping is rude and unnecessary. Be aware of your surroundings to avoid being a bother to people in your vicinity.

Rule #3: Think about where you are. Ask yourself: "Will this bother the people around me?" before you make a phone call or answer one. If you are in a place where nobody is talking, assume that the people there want to keep it that way.

Rule #4: If there is a sign that states, "no cell phones permitted" that is exactly what it means. There is a reason for the posting of these signs and makes it even ruder to carry on a phone conversation in an establishment containing a "no cell phones permitted" sign.

Rule #5: Think about who's near you. If you are planning to talk about someone on the phone be sure they are not within hearing distance of your conversation. When you say something like, "I'll tell you later", it is obvious to the person near you that you are talking about them.

Rule #6: Are you already talking with someone? Do not answer your phone, text or use Facebook while you are in the middle of a face-to-face conversation with someone else. It is beyond rude to answer your phone when you are talking to someone else. It is pretty much saying, "I have more important people to talk to."

Rule #7: STOP any phone calls while you are driving. No conversation is important enough to put yourself and others in that danger. If it is important pull over and answer your phone. And equally important, **don't try to text and drive**: it's hard to finish the text before the light turns green, holding everyone behind you up OR you end up trying to insanely text and drive.

Rule #8: Ringtones can be more obnoxious than the people on the phone. If you are not going to answer your phone, hit the quiet button so that it stops ringing. People do not want to hear the "Pussycat Dolls" blasting from your phone.

Rule #9: YUK! Do not bring your phone into the bathroom. That is extremely unsanitary and is absolutely not the place for a conversation, ewe!

Rule #10: Close with class. If you want to end a conversation with someone, simply tell him or her that you have to go. There is no need for the tacky "candy wrapper" trick in order to pretend you are losing the connection. It is also quite obvious that a candy wrapper is not static; people are smarter than they are given credit for.

Rule # 11: Multiple conversations. If you are having a phone conversation, do not text someone else in the middle of it. If you have more important things to do, then end the conversation because texting while on the phone with someone else is rude. The person on the other end can tell! They can either hear the clicking of the keys or, in many cases, you just agree to whatever the person is saying because you are distracted.

Rule # 12: Museums, movie theaters, and plays are meant to be silent activities. Texting or talking in this situation is never acceptable. Refrain from using your phone in these instances at all costs. If you must, then it is polite to step out of the room to take your phone call or answer the text message. In addition, the light from your phone is distracting to everyone around you in a dark theater.

Voicemail

The rules of voicemail are very important because most people see it as an inconvenience more than anything else. These days, leaving a voicemail message is probably the most annoying thing a human being could ever do to other human beings by means of communication [other than the list of things we noted above].

What is Voicemail?

Well, voicemail was created to allow thoughts or ideas to be received by someone else even when they are not able pick up their phone. Here is the situation: you call someone. They did not pick up the phone. You want to warn them about a surprise pop quiz the teacher is giving because your friend planned on skipping class. If the person does not go to class, the score for the quiz will be zero, impacting an already poor grade. What do you do? You leave a voicemail. You say what you would have said to the person if they actually picked up and send the message to their phone. This way when they pick up their phone they will hear the message you sent. Voicemail allows for a deeper means of communicating indirectly and seems to have solved the problem if someone is unable to pick up their phone.

Problems with Voicemail

There are a few problems with voicemail that would make someone go anti-voicemail which are:

☐ **The long pause.** It is incredibly annoying to have to wait until the beep comes along before leaving the message. First, you have to wait for those annoying five rings to go through before the phone even sends you to voicemail. Then, you are told that the person you are calling is unavailable (as if you could not have already guessed that). Finally, you are told what to do after you finish leaving the message when just hanging up usually is enough.

□ **People may not even listen to them.** There is no guarantee that someone has even heard the message, or realized they had a missed call. Voice messages become old and nobody wants to take the time to listen an old message. Or maybe someone is angry at you and does not want to hear anything you have to say, which ironically could explain the situation and solve everything. There are many instances when a voicemail just does not go through.

□ **Poor setup.** Just the way you experience the voicemail setup makes you despise it. For starters, having to use a password every time you access your voicemail quickly becomes redundant and annoying. Next, you are forced to listen to the voicemails starting from the first one, which can become a hassle if you have many messages and want to only hear the last one. Then there is the choice of deleting the messages; although this may seem like a savior at this point, it completely becomes the opposite as you are told over and over if you want a message deleted followed by the same voice telling you it was deleted. It's never easy to get around the use of voicemail.

□ **They clutter quickly.** If you are not up-to-date with the calls you miss, your voicemail could become very large. Listening to all of the messages could take an incredibly long time. You would have no choice but to listen to all to determine their importance, which could become more of a chore than anything else.

Why Voicemail Ultimately

Failed This Generation

Ultimately, the idea of voicemail failed because of the way the world has evolved socially. With tools such as texting, Facebook, and other means of communicating in the palm of our hands, voicemail never had a chance with our generation. Ideas have become much easier to get across because now we can simply let the people that are the hardest to reach know anything we want by just texting them or contacting them through social networks like Twitter© or Facebook. Younger generations typically say "no" to the use of voicemail, although older members of our society still use this method of communication.

Safety

Safety when using cell phones is a very serious issue and should not be taken lightly, especially with all of the problems it could inevitably lead to. Now we could go on to explain the issues of talking by just listing them out clear and dry, but that is not really fun from either your perspective or ours. So real-life situations will be applied because it makes it a little more interesting for all of us.

Talking and Driving Safely

You are driving home from a party and a friend calls you to talk about what happened. This person just happens to be your best friend, and someone you would not want to ignore. So you pick up the phone and start having a conversation. In the meantime, all of the cars in front of you have suddenly stopped and you hardly notice. One quick glance and the next thing you know, you are waking up in the hospital not knowing what happened.

It's unfortunate to say, but this happens more often than expected and usually the outcomes are not so good. Whether it is talking on the phone, texting, or checking your Facebook status, please remember that more important things come before it all. One call is never so important that it cannot wait. This rings true according to a study done by the National Safety Council this year that concluded 28% of traffic accidents occur when people talk on cell phones or send text messages while driving.

Watch What You Say

You and a friend had a fight and do not want to talk to each other anymore. You want to play a prank on her because you believe that she deserves it. So you call her house and her mom is the one who picks up the phone. You say that there is a bomb in the house; it is ready to explode and then hang up. The next day, the police track the call to you and come knocking at your door. The next thing you know, your hands are chained together and you are in the back seat of a police car.

Most people do not plan out what they say and the consequences behind it. Anything you say can have a negative effect and hiding behind a phone is no different. You are responsible for what you say in this world regardless of how it is communicated. Causing distress any way, shape, or form can be very dangerous as it alarms a group of people which could then lead to traumatic disturbances in their lives. So think before you speak, and be careful of what you say.

Consideration

Every situation is different and no two people are the same; therefore, it would be impossible to go on and explain every situation and the negative outcomes due to talking on our phones. Just because something was not discussed in this situation does not mean it is automatically safe. Be mindful of where you are when using technology and be aware of making the right choices even if they may not always be obvious.

Pros and Cons of Less Talking, More Texting

Pros

- Speed of communication is greatly increased

- Since scheduling a time to meet and talk is sometimes overwhelming and difficult due to people's busy work or school schedules—the texting alternative for brief and precise conversations is an excellent tool to take advantage of when you are in a hurry.

- Most of us are able to text at a much faster rate than we can talk.

□ **Less time can be spent on outward appearance**

- We are becoming more and more technologically advanced as a society to the point where face to face conversation is no longer the norm (webcams and phone conversations are starting to take precedence), so we can spend less time making ourselves look presentable. This, in turn, leaves us some free time for other areas of our lives. Most people spend a lot of time carefully picking out an outfit to wear to an interview or meeting in order to make a good first impression, but if that meeting were to instead take place on-line, a businessman or woman could theoretically be wearing slippers and nobody would even know.

□ **More convenient and discreet**

- Nobody likes the people at the grocery store or bank blabbing away on their cell phones; text messages are a perfect alternative since they are virtually silent and allow private conversations to take place in public areas like a library, waiting room, or elevator.

- Texting is more convenient. With phone conversations, both parties need to be free to chat on the phone, whereas with texting, only one-person needs free hands to text another. They can rest assured by the little check mark icon on most phones that their intended receiver did indeed get their message. The individual who received the text in their inbox can then answer at his or her own convenience as opposed to playing phone tag.

□ **Fast and efficient**

- Text messages are expected to be short and concise.

- You can abbreviate and be to the point without coming across as rude or inconsiderate.

- You cannot say, "Movies at 7:00pm with Kate, Ryan, and Sandra. Be there!" on the phone or in person to a friend and then hang up or immediately walk away; however, with texting as an option, the above comment is appropriate.

□ **Gives you time to choose your words carefully**

- When texting you automatically have the chance to erase a word and replace it with a better one.

- Since we are able to proofread and edit texts before sending them, we can avoid hurting people's feelings or arguments and discuss how we really feel.

- Whereas fighting via texting is unconventional, it is useful for those of us with heated tempers since we do not have to be in the same room with the person during the altercation, but we can still express what is bothering us by physically typing it out.

- This way, each party gets to discuss his or her side of a story whereas in a phone-call argument or face-to-face argument emotions can often be too overwhelming and the two participants may end up getting physical or walking away from each other—neither of which are healthy alternatives or problem solving tactics.

- With the ability to focus on what is really bothering you, texting becomes a tool to help you communicate better with a person you have trouble thinking on your feet with when you are face to face.

☐ **Eliminates the need for competency in grammar/spelling**

- Can't spell tomorrow? No problem. Not sure if definitely has two "f's"? Once again—this is not a problem. Can't remember if it's to or too? Still, this is not an issue (2mrw, 2morrow, def., and 2 are all suitable substitutes!).

- People who have poor grammar skills need not worry since it is perfectly acceptable for people to sound like illiterate morons; abbreviations are actually highly encouraged.

- Also, for the texts that you feel need to be spelled out—i.e. ones to an authority figure or an elder, there is T9 or Word for that! These are mini programs that are built into most standard cell phones that try and auto-complete the word you have started to spell by guessing among popular used words in the English language.

☐ **Easily retrievable**

- We normally do not record our telephone call exchanges or day-to-day conversations with people, but with text messages, we can save them and refer to them whenever we please.

- All messages can be "locked" or saved so that if there is important information in the message such as an upcoming event location, date, or time or even another person's phone number, fax number or address, we can store that information without having to memorize it.

☐ **Makes mass communication simpler and more commonplace**

- Instead of calling a phone chain of one hundred and fifty employees or students about a change in a meeting or event, several ten-person-at-a-time texts can be sent out to cut down on small talk and wasted time in relaying the memo (at this printing, most cell phones can send one message to a maximum of ten recipients at once).

☐ **Cost efficient**

- Especially when someone is communicating internationally, text messages are relatively inexpensive when compared to charges that rack up by the minute during phone conversations.

Cons

☐ **Inability for young people and adults to converse in person**

- With texting (along with other technology such as instant messaging, Facebook, e-mails, faxes, phone calls, and video chatting), face to face conversation is slowly dwindling and becoming less imperative.

- A frequent texter may lose his or her ability to pick up social cues and read body language and facial expressions if they never have person-to-person interaction.

- If nobody is leaving his or her house or office to talk to someone, life may become impersonal and dry.

- Talking for longer periods in person helps to develop intimacy. Texting lessens the need for face to face contact and reduces opportunities for practicing the art of intimate communication.

☐ **Lack of emotion/inability to express ones feeling verbally**

- With so many teenagers and young people texting as their primary means of communication, it becomes more difficult for such groups to express out loud how they feel emotionally.

- Since they are not used to saying how they feel, this

could build stress. Without healthy ways to reduce stress such as expressing feelings, someone may snap after not talking to another person or confiding in someone they trust for so long.

☐ **Ineffective multi-tasking that leads to a lack of productivity and a lack of focus**

- With so many stimuli out there between computers, video games, and now cell phones, students can be on Facebook, talking on the phone, texting, watching TV, and typing a paper all simultaneously.

- This type of multi-tasking is highly inefficient since we cannot focus our attention on that many things effectively, according to researchers.

- Cell phones are instigators for procrastination and it is best if they are silenced and away from designated study areas.

☐ **Dangerous everyday situations**

- Since texting is as common as talking, people think they can text while doing any other thing they want including driving, walking, and exercising.

- Texting cannot be paired safely with activities that require both hands! People who text a lot have a very hard time putting their cell phones down or away. This is a serious and little known side effect: when does it become an addiction?

☐ **Less literacy and proper grammar**

- Texting language has spilled over from messages into student's school work and adult's e-mails. This is simply unacceptable and highly unprofessional.

- Kids need to know how to spell "because," "without," "tomorrow," "anyways," and other commonly abbreviated words in order to function in the workplace.

- Those that cannot send a professional and articulate letter or e-mail are not going to get a job over someone who can.

- Abbreviations where they do not belong make the user seem dumb.

□ **Unintended viewing**

- Sending someone a text message could be as public as writing your words on a billboard.

- Once the send button has been pressed the message goes to another person's phone, which can be read by anyone at anytime, anywhere.

- Unless you trust the receiver one hundred percent and know that he or she carries their phone on them at all times, you should never send anything too personal that could be accidentally or purposefully read by someone else.

- You never know if the person you are texting is alone and therefore you have no idea who else is reading the messages you are sending.

- There is ambiguity as far as how private your message actually is.

□ **Miscommunication**

- Without seeing the person's reaction who received your message, you have no way of knowing how they interpreted the words and sentences you sent.

- Tone of voice and body language are eliminated during texting and so you may find that you are misunderstood by others or that your messages are unclear.

□ **Missing out on new opportunities**

- Some people use their cell phones as a crutch and text old friends in new situations where they are nervous or not paying attention, in the hopes to minimize anxiety or avoid the new situation they are currently in.

- Checking one's cell phone for messages or the time during an interaction with a person or people in a room can be rude and distracting. It is also rude in the dark of a movie theatre or at a family dinner celebration.

- In order to meet new people and maintain relationships with people you have already met, it is important to focus on enjoying the time with those around you instead of texting other friends at the same time.

□ **Addicting behavior that can ultimately lead to wasted time**

- If a person communicates socially through text messages, then he or she is going to feel naked without their cell phone and be compelled to respond to it whenever it beeps or alerts them that they have a message.

- A frequent texter may also develop discomfort or anxiety when it comes to actually calling another person to have a conversation.

- The immediate gratification our generation gets from receiving a response to a text message is enough to make us forget everything else and can lead to a massive amount of wasted time.

- That immediate gratification has also led us to expect faster responses in other areas of our lives like return e-mails from teachers, return phone calls when we are forced to leave a message, immediate service at a restaurant or doctor's office, and finally next day delivery when we order something on line when it can't be instantly downloaded. We have less impulse control.

- When having a face-to-face conversation, two people have a verbal exchange—and then it is over. If we are texting a person, however, there is no real closure or clear end to the conversation.

- Those of us who text numerous people at the same time can find ourselves doing nothing but texting for hours to keep up with everyone's responses. This is precious time that could have been allotted to more productive activities.

Using a cell phone for talking will soon be obsolete because texting is a prominent feature of the cell phone. The next section will explain the details of texting.

Chapter 2

The Convenience Of Texting

The 21st Century has brought in a whole world full of new technology, presenting an entirely new medium of communication: text messaging. Text messaging was originally intended for the purpose of sending quick, brief messages on the go, from your cellphone. Now, text messaging has become such a critical tool in young people's social life. Texting is now such an art that it has its own terms and abbreviations used exclusively in texts. When texting, everything from the time elapsed in between messages, the length of texts, underlying innuendos, and even the tone interpreted through the messages are all taken into consideration, making texting an established and necessarily critical skill.

Relationships, even romantic ones, are now formed through texting. However, text messaging presents complications. Friendships as well as romantic relationships are often destroyed because of the misunderstandings that come with texting. Text messaging is no longer a way to conveniently communicate, when e-mail or speaking on the phone are not an option. Texting is a huge part of how young people, and even some adults, communicate. Text messaging is the outlet many young people express themselves through and serves a crucial foundation in their social lives. In this chapter we hope to lay out different scenarios and talk about the rules that govern those situations.

Rules and Behaviors for Guys

Texting has become such a vital part of most people's lives today. It is impossible to go anywhere without seeing someone on their phone texting. It is just the new fad that everyone is following.

With texting comes some important rules and behaviors for guys to follow, whether they know it or not. These can be split up in different categories. There are certain rules and behaviors when texting a girl, but while texting a guy, the rules change.

The rules for texting girls can be quite difficult to follow sometimes. The most important thing to remember is to keep the conversation going. It is very difficult to text a girl without saying things to keep the liveliness of the conversation. Even though it may be difficult to think of new things to ask or to say, it is vital that it be done. Having the conversation die out is something that no one wants. To keep the conversation going, ask questions about what the other person did today or questions about the person's interests (movies, color, favorite food). Next, it is important to not curse. It is very uncommon for girls to enjoy being cursed at. Even though this is a rule, it is not one of the most important rules to remember. Another rule is to not give one word answers. One word answers can be very frustrating to deal with. One word answers can stop the conversation from flowing easily, which is what you are striving towards. Also, giving one word answers can make it seem like you are not interested in talking to them. Obviously, for this reason, one word answers are not good. These are the most important rules and behaviors to remember when talking to a girl via text message.

On the other hand, rules for texting another guy are fairly simple. One word answers are acceptable, unless you are in a deep conversation with a friend about your problems. Most guys do not mind getting one word answers because it isn't a big deal for them. They usually just ask a question about plans for that night or day and a quick and easy response is all that is needed really. Basically, texting rules for guys to guys are much simpler than texting rules for guys to girls.

There are a few rules that can overlap, though. For instance, answering within a reasonable amount of time is always greatly appreciated. It is usually a pet peeve for both sexes when the person you are texting takes a long time (more than 10 minutes) to respond. A response should occur within a short period of time (less than 5 minutes), unless something dramatic happens. Also, actually answering the text message applies to both males and females. If someone texts you, do not ignore it; show some respect and answer them back.

Overall, there are many rules to be aware of when texting. Girls have certain rules to follow as well as guys having rules.

Although, there are some rules that can overlap and apply to both sexes. While texting someone, it is extremely important to remember what rules you should be following depending on who you are texting.

Rules and Behaviors for Girls

Before talking about the rules for girls see how you answer these simple questions.

1. Always checking to see if you have a new text?

2. Hate when the person you're talking to does not know how to keep the conversation going?

3. Hate when the guy you're interested in takes a long time to respond?

4. Find yourself texting in awkward situations to seem occupied?

5. Never going anywhere without your phone?

6. Sleep with your phone?

If you answered yes to three or more of these questions, then you are a typical girl texter. There are many rules and behaviors that most girls follow in the art of texting.

First things first: when it comes to the opposite sex, the girl usually does not initiate the conversation, unless you two are dating. It is up to the guy to make the effort and show how much he is interested. By texting first, a girl seems too eager and vulnerable. A girl must always remember that it is better to play hard to get. A guy will not find it attractive if a girl is texting him nonstop and making him feel overwhelmed. If he wants to talk to you, he will. Next, when the person of interest finally does answer, do not respond right away. Give it a few minutes to make it seem that you are not sitting right next to your phone (even though you are). Then it is up to each individual to make the conversation last. Girls are much better at making something out of nothing when it comes to talking to a guy, like ask about past relationships. When a girl sees the conversation starting to die, they then take control of the situation by starting a brand new topic or ending the conversation completely. These behaviors come naturally to most girls.

Girls are the ultimate multi-taskers. We can incorporate texting into almost any activity. That is why girls become frustrated when someone does not answer in a reasonable amount of time. If girls can do their homework, be on Facebook, watch television, have a conversation with their roommate, and text someone back, there is no excuse for why the person on the other end is not answering. Guys, if you are doing something and know you will not be able to respond, just let the girl know and all problems will be avoided. Most importantly, never leave a girl hanging in the middle of a conversation; she will not be a happy texter.

One thing you will never find is a girl without her cell phone. A girl's cell phone is her life line and best friend. A cell phone will never let a girl down, when everyone else already has. Do not take it upon yourself to start browsing through her phone to see who she has been texting and calling. This will most definitely lead to a fight. If a girl wants to let you into her cellular world, then she will on her own terms. Especially all you boyfriends out there: do not look through your girlfriend's phone. Not because she is hiding something from you, but more just out of respect for her privacy. One thing that parents should remember is that going through your daughter's cell phone to find out information just looks as if you do not trust her and will lead to confrontation. If you are a parent and have a question or concern, just ask instead of trying to be sneaky about it.

Courtesy of Dating

When texting a person of the opposite sex (or the same sex) there are many unwritten rules that should be known to everyone about how to keep the relationship alive. When you text someone when you first start talking and you want to get to know them, you should always play it cool. Girls, do not text guys first. It gives most guys the idea that you're either desperate or just plain annoying, and no girl wants to be known as that! Smiley faces are becoming more popular through texting and give more personality to the conversation. Overused smiley faces are a bit much when you first start trying to form a relationship with someone. There is a max on how many times you should text somebody after they do not respond to your first text. After one text, give the response time a decent amount. The person receiving the text may be busy, or not near their phone right away. Later in the day, you can text them again. If the person STILL does not respond to you, it's a lost cause. It stinks for both parties, but there are plenty other fish in the sea so just move on and don't take it personally! Whether you are dating somebody or just beginning to talk, something that is universally said by a girl to a guy is, "okay I'll talk to you in the morning..." When a girl says that, 99% of the time this means that she is expecting you to text her. Some guys believe that girls expect too much, and some guys even think that they are being annoying. But a girl would not say that if she really did not want you to text her! (TRUST US!)

The rules change, however, once you are actually dating somebody. Once you begin to date someone, there is a trust and understanding that is formed already. Some of the rules still do apply though. Do not bombard the other person with duplicate messages and overused smiley faces. A "Good morning" text from either sex is very reasonable and not at all too demanding. When you're with your boyfriend or girlfriend, it may cause some problems if you are texting the entire time that you are with them. Taking a boyfriend or girlfriend's phone is a BIG insult; not because the person is trying to hide something and be sneaky about it, it is just the privacy aspect of it. In a relationship you are supposed to be open with each other. But on the other hand, you both need your own privacy and a phone is a big source of privacy. When it is taken away and looked through it may not end pretty. Before using or looking at someone's phone, always ask first; it just shows respect and demonstrates trust.

Fighting via text is also another issue that should be touched upon. This does not only go for relationships, but it also goes for friendships as well. Our generation now argues a lot through technology. There is rarely any face-to-face confrontation anymore. When fighting with a girlfriend or boyfriend through a text, there are two important things that occur. One, that the receiver has enough time to think of a response or even in some cases a lie to prevent further fighting. Second, when boyfriends and girlfriends argue through texting, some people can be meaner through a text. But, if they see someone's reaction when something is said, they may regret saying it because of the other person's reaction. For example, say I wanted to tell James that I wanted to break up with him. I do this through a text message because if I see James get very upset in person, I may not actually end up going through with it.

Texting has completely changed people's relationships. There have been stories about people who have texted for a year, and never actually met each other, but were madly in love. People are beginning to believe that texting is the new talking. Is it? Is texting ruining relationships? Or is it making relationships stronger? We'll let you decide.

Sexting

Sexting, in general, is when one expresses to someone their sexual desires through a text message. However, sexting is much more complex. It takes a decent amount of time to pass and trust to form before sending these intimate little messages. Sexting begins with very innocent roots. It all starts with the "hey what's up" which is followed by some more superficial, harmless conversation. Eventually, you will find that you are expecting or sending texts, to this special someone, routinely. You begin to get an idea of each other's schedules, as well as learn the about the average amount of time it takes them to text back. Texts become lengthier and deeper conversations start to transcend. Many would agree that the text that confirms mutual attraction is the "good morning" text. The "good morning" text (especially if being sent daily) proves that you are the first thing on this person's mind. "Good morning" texts have more significance than "goodnight" texts because "goodnight" can simply mean "I have to go"-ending conversation entirely.

After the "good morning" text, one will now find themselves getting butterflies in their stomach every time their phone lights up with a text from their new crush. Now is when the compliment phase begins. The compliment phase is when the texting pair starts to point out each other's best physical attributes. This is fun and flattering for both parties. Emoticons (smiley faces used in texts which are arranged by different letters and symbols) are key to flirty/sexy texts. The "wink-y" face (a semi-colon and a right parenthesis) can be used to imply attraction and confirm any sexual undertones in texts. Once each other's sexiest features are recognized, the braver texter will get the nerve to send the "we should hang out sometime" text. This text can reasonably be translated to mean "we should start hooking up." Regardless, the recipient will gladly agree and arrangements to see each other in person will be made. Either before or after the texters hook up for the first time, the true sexting will start.

Sexting is risqué and compelling, but has to be done with caution and respect. Only after trust has been formed between the texters can sexting take place. One must have concise timing when initiating the first "sext." Sexts cannot be sent too soon into the texting relationship. The first sender of a sext must be confident that their message will be well received by their significant texter. The first sext can begin with a compliment (as discussed earlier) and then can be followed by pointing out how good they must look in little to no clothing- for example. Soon the more graphic texts entailing detailed, sexual desires will be exchanged. Imagination and the expression of one's own sexual fantasies are encouraged for an optimal sexting experience. Sexting allows for a fun and erotic connection in the comfort of...wherever you may be texting.

Sexting is definitely exciting, but can be risky. Beware that sexts can serve as incriminating material and be devastating to one's reputation. It is advised that after sexting, all sexts are deleted. You do not want to risk your mom, friend, younger sibling, or ANYBODY to read your most personal, sexual desires. If you are someone who enjoys reminiscing and wants to keep your sexts forever, make a point of having a password on your phone, so it is accessible to you only. Other things to keep in mind when sexting: always double check who you are about to send that suggestive text to- god forbid you accidentally send it to your dad, the annoying creeper who is always hitting on you, your boss, your grandma, etc. The consequences of sending a sext to the wrong person can range from being mildly awkward to finding yourself in some serious trouble. Regardless, sending a sext to the wrong person is embarrassing. The worst possible, most devastating scenario in sexting would pertain to the naked picture. This usually pertains to girls, but is not limited to girls. Often times, regardless of relationship status, a girl will send a boy pictures of herself in revealing clothing, parts of her body, or completely naked. This is either done upon request or voluntarily. The consequences of sending these pictures can be devastating: the recipient only has to forward the picture to one person before hundreds have seen it. Once the picture has been forwarded, it is out in the world and can never be taken back. This is humiliating for the sender, as well as damaging to their reputation. They have compromised their privacy and pride just because of pressure or succumbing to the heat of the moment. Although sexting is fun and can be relatively innocent, sending a picture of yourself, or even receiving one, is not recommended because the possible consequences outweigh the initial rush.

Sexting is a relatively new term, but is extremely familiar to young people. Sexting is not exclusive to any one age group, though. Sexting is done discreetly but is done a lot more frequently than people would assume. Even the most innocent, conservative people are sexting. Sexting provides an outlet for people to release sexual tension, letting go of inhibitions that they otherwise wouldn't express in person. Sexting is convenient and can arguably be done anywhere, anytime. Sexting, like any risqué act, does not go without any consequences. Carelessness or poor judgment may result in sexts coming back to haunt you. So be wise and cautious when sexting, but definitely make sure to be creative and most of all, enjoy!

Response Times for Different Situations

For all the new texters out there, this section might be a little overwhelming at first, but just keep reading and you will learn all you need to know. You may not think that when someone texts you, there is a certain etiquette for response times, but there certainly is, and it is very important. It is common to believe that no matter what the situation is, you should just text when you are not busy, but in this day in age, that is not the case. There are different response times for multiple situations. It is impossible to list them all, but this will give you an overview that will help you be able to calculate the response time for any situation you may be in.

The first and most important aspect that comes into play when figuring out response time is who you are texting. To make understanding this a little easier, let's break down response times by certain people. Knowing how long to wait in between texts according to whom you are texting is one of the most essential things you must master in order to become a great texter.

Texting A Friend

Response times for texting a friend is the simplest of all situations. It may vary according to how good of a friend the person is, but for the most part it is all the same. If a friend texts you, you should never intentionally keep them waiting. If your friend texts you and you are not doing anything, you should answer within five to ten minutes. By answering in this time, your friend will know that you are able to talk and the conversation will continue smoothly.

If you become busy in the middle of a text conversation, you should tell the person you have to go. Do not just stop answering them! They will feel like they are being ignored and most likely assume you are mad at them or something is wrong. This may lead to tension in your relationship until it is straightened out.

If you are busy when you first receive a text from a friend and you cannot answer, you should make sure that when you do get the chance to text them back, you start the conversation with an apology that you could not get back to them sooner. Normally if you are not available for texting for a long period of time you will tell people beforehand, by text or Facebook and/or reply quickly that you can't talk. If you do this, there usually will be no problems whatsoever. Now, that wasn't so bad, was it? It gets a little more complicated, but remember, when you are done reading; you will be an awesome texter!

Texting A Boyfriend/Girlfriend

As silly as it may sound, when you are in a relationship, texting becomes a huge deal. It can even cause fights between you and your boyfriend/girlfriend, but if you know the rules on response times, you're in the clear. If your boyfriend/girlfriend texts you, you should answer them right away, or as soon as you can. During a conversation, it is important that you do not wait more than five minutes or so before texting the other person back. Responding quickly is key to making the conversation flow without any problems. When you do not answer in a timely manner, it is likely that the other person will get angry or upset. If you do not answer and they text you again, you are in trouble! Whatever you do, try to avoid allowing that to happen. It will cause unnecessary arguments.

The only time that you should not respond to your significant others' text message quickly is when you are already in an argument, or when they have done something wrong. If your boyfriend/girlfriend has done something wrong and they are texting you to try to make it better, don't answer right away, let them sweat it out a little bit.

If you understand everything up to this point, you're a fast learner and you should have no problems with texting at all. Keep on reading, you're doing great!

Texting Someone New

When you first meet someone, if you hit it off, you may exchange numbers, and in a day or two you will start talking to this person regularly. This section is the opposite of the first two, because these types of cases you do not want to answer right away. If a new person sends you a text, you want to wait at least 15 – 20 minutes before you answer their first message. After that, you should wait between 10 and 15 minutes in between each text. Why is this different then the other situations? Because you want this person to think that you are busy. If you are taking a while to answer, chances are they will be wondering what you're doing that is making you not answer them right away. This wondering leads to mysteriousness which is a great thing to have at the beginning of a relationship. Being mysterious often sparks interest and may make the other person want to get to know you better. See? Knowing response times can totally work to your benefit.

Texting Your Parents

Response time for texting your parents is very straightforward and simple. Answer them right away! No ifs, ands or buts; do not make them wait. For some strange reason, parents are prone to being neurotic, and crazy things will run through their minds if you do not answer them. If you do not answer your parents' texts quickly, it is likely they will think one or all of the following things: you are hurt, something is wrong; you are doing something bad, you are somewhere that you should not be, or you are in some kind of trouble. The only way to avoid this is to answer right away. If you answer right away the conversation will be short and smooth, which is just what you want, isn't it?

Well, there you go, now you know the main points of response times for different situations. Think you got it all down? Only time will tell! If you are still completely unsure and overwhelmed by the information, just try to remember the following tips. They are the basics of the basics and if nothing else helped, these guidelines surely will.

When thinking of how long to wait between text messages,

first consider who you are texting. That is the most important; it all depends on the person.

□ Next, think about the importance of the conversation. If it is high priority answer right away.

□ If you are trying to maintain a relationship with a significant other, do not let too much time pass between text messages. You want to avoid possible arguments as much as possible, and that is a simple way to do it.

□ When it comes to parents, there is NO reason not to answer right away! They will never accept you not answering right away and you should not give them a reason to get mad at you.

□ If you are really unsure about how long to wait, choose a midpoint. Ten minutes is a good midpoint, so if you really do not know when to answer, give it ten minutes.

Conveying Tone via Text Message

When communicating with someone via text message, it is important to consider your tone. Your attitude is not the only component of tone; it is made up of many different parts that must be taken into account when talking to someone. This section will go over the many ways to communicate your feelings with ease.

One-Word/One-Letter Text Messages

It is very important to understand what most people think about one-word and one-letter text messages. One-word text messages are usually a sign of anger. When you send a long text message and someone replies with "Okay" or something to that affect, you are bound to be dissatisfied. Wouldn't you want a long response in return? That dissatisfaction is more than likely the effect your conversation partner is hoping for. Take into consideration that this person may be at work or just too busy to respond correctly to a text message. However, this person may simply be a "bad texter". He/she may simply not understand the negative connotation associated with one-word texts. Bad texters should be made aware of the one-word rule. If you get a lot of one word answers maybe it is time to end the conversation. Some words to avoid when responding to text messages are:

- Okay
- Alright
- Uh-huh
- No

- Nope
- Nah
- Yeah
- Ok

You should make sure to never leave these words alone in text messages. However, the mother of all bad text messages has not even been mentioned yet, and that text is "K". Saying only "K" instead of OK it is definitely a no-no. When someone says "K" and only "K" in a text message, it means he/she is definitely angry with you. However, sometimes bad texters and older people (such as parents!) are unaware of this rule. If this is the case, you must make them aware of the K rule! It is never acceptable to break this rule, ever.

Multiple/Capital Letters

Some people, usually preteen and teenage girls, use repeating letters and mix capital letters in the middle of words in their text messages. There are different kinds of repeating letters. There are mild cases such as saying "sooooo" instead of "so", more moderate cases such as saying "I knowwwww" instead of "I know", and finally extreme examples such as repeating multiple letters in one word ("tooootallyyyyy") or repeating letters in multiple words in a sentence ("I likeee don't even knowwwww"). Mixing capital letters is another style of text messaging. An example of this would be "HaHa ThAt Is SoOoOo FuNnY". (Yes, everyone knows this looks ridiculous, but people do it.)

This is a very unprofessional way of texting and it should be stressed that although it is sometimes okay to use this form of texting with friends, it is unacceptable to use with people in a professional setting such as employers, coworkers, and other people who you communicate with in a professional manner. Another point to be made is that repeating letters and mixing capital and lowercase letters is for younger people. That is not to say that people in their 20's and 30's can't utilize these styles, but it is preferred that your texting style ages and becomes more professional as you do.

Avoiding Sarcasm

It is important to keep sarcasm in mind when having a conversation with someone via text message. There is a huge difference between talking to someone face-to-face and talking to them via text message. Sarcasm has a lot to do with voice inflection and facial gestures, neither of which can be understood in a text message. Therefore, sarcasm can be easily lost in a text message; it can either go right over someone's head or easily hurt them even if you did not intend to. For that reason, sarcasm should be avoided whenever possible. If you are a naturally very sarcastic person, this might be hard for you. You can use sarcasm sometimes if you feel that your conversation partner will understand that you are being sarcastic. However, it is generally not a good idea.

Be Honest

It is very important to be honest about your feelings when talking via text message. If someone upsets you or makes you angry, don't tell them you are okay if you are not. Like sarcasm, hurt is not transferred well via text message. When someone does something that bothers you and you say you are fine, it is likely that they will believe you are really fine because they can't hear the hurt in your voice. It is perfectly acceptable to tell someone when something is bothering you, if you truly want them to know. If you want them to simply "figure it out" you can hint to them using the one-word and one-letter texts described earlier in the chapter, but if you honestly do not want someone to know they have hurt you, then by all means tell them you are fine. If you convey hurt feelings via text message and it comes to the start of a fight, it is not recommended that you fight via text message. It is impersonal to do so, and not fair considering that it gives both parties time to pick and choose their words wisely and decide what they want to say. In a fight over the phone or in person, it is much more honest because people must say the first thing that comes to mind instead of thinking about what they want to say in order to manipulate the argument in their favor. That being said, it is preferred that you call someone or meet them in person to have a serious argument instead of text messaging them. Being thoughtful about arguing is a good thing and you can carefully choose your words to not be hostile or say something you regret or are extremely painful to the other person. However, using the time to come up with lies or manipulations is just as bad as calling name and saying the regretful or painful things.

Know Your Audience!

Knowing your audience is probably the most important component to conveying your tone via text message. Just like you would adjust the way you speak when talking to a friend and then talking to a professor, you must adjust your texting style to accommodate those with differing styles than your own. This is not only a professional move, but also a simple courtesy. You don't want anyone to misunderstand you and you don't want to misunderstand anyone else. It is an unspoken rule that when two texters with differing styles text one another, they need to make slight changes in their texting style to make the conversation flow better and to make it easier to get points across without misconstruing either person's point. It is crucial to understand not only your conversation partner's texting style, but also their personality in general in order to have a successful and satisfying conversation.

Duplicate Messages

What is it?

Duplicate messaging is when the sender of a text sends either the same message twice in a row before the receiver responds to the message, or a completely new text. The sender does this when they do not get a response.

When Does It Occur?

This type of message occurs when the sender of a text does not get a response from the receiver. The sender may send it because the receiver has not responded fast enough or they just didn't respond all together. The sender may become very impatient because the receiver of the text message is not responding. When the receiver does not answer it could mean that the receiver does not want to answer or they are just busy with something else.

Is it Okay or Not?

It depends. If the person you are texting is a boyfriend or girlfriend, then a duplicate message is okay. However, if you are texting someone you are "talking" to then a duplicate message is not okay! Duplicate messages are often very annoying. They will answer under their own conditions, and if they want to. If they do not answer, then they probably do not want to talk to you, or they are just busy doing something else. Another case where this is acceptable is if it is a friend or family member. These instances are the cases where a duplicate message is okay. If you text them something but have sent the message too quickly then a duplicate message is acceptable. However, just be careful not to do it too often, and always keep in mind the person you are about to send the messages to!

Deleting your messages

Several people have the tendency to erase their messages, whether they are text messages, picture messages, or emails on their phone. There are both positive and negative aspects of doing so, which are outlined in the following reasons:

- Erasing messages to save space: Some people erase their messages to save space of their cell phones. When an inbox gets too full, you are required to erase messages in order to receive new ones.

- Erasing messages to secure financial information: Some people may have their financial information sent to them through text messages or email. They may erase these messages to ensure that nobody else will be able to get a hold of it.

- Erasing messages out of embarrassment: Some people erase messages that they feel are embarrassing. This can include either messages sent or received that others would find unusual. Erasing these embarrassing messages ensures the user that nobody else will see them, and that they won't be used against user.

- Erasing messages that contain troublesome information: Some people erase their messages that contain information about something they have done illegally. This can include talk about drug/alcohol use, stealing, violence, etc. People often erase their messages so that if it got in the hands of someone else, they would not be sanctioned for anything that they have disclosed.

- Erasing messages to hide something positive: Some people erase their messages if they are planning something for somebody else, and do not want them to know. For example, if someone is throwing a surprise party for their friend, they will delete their messages to make sure it stays a secret.

- Erasing messages to hide something negative: An example of this would be if a person was cheating on a girlfriend, and did not want her finding out. He would erase any messages that would reveal this information.

☐ Erasing messages to forget a conversation: Some people erase messages that they want to forget about. For instance, if you are fighting with a friend through texting, you may erase your messages in order to help you forget that it happened.

☐ Erasing messages that are unimportant: Some people erase messages that serve no significance to them. For instance, some people may erase an email on their phone that does not pertain to them.

☐ Erasing picture messages: Some people send pictures that are only meant for one person to see, or receive pictures that they do not want others to see. They may erase these messages to keep the pictures private.

These are some of the general reasons why people tend to erase messages on their cell phones. As you can see, there are both positive and negative reasons for doing so. Generally, people who erase their messages do it for privacy. It is important to keep certain information to yourself, as well as have the ability to keep some information hidden from others. However, it is important to keep in mind that even though you may erase your messages, they can usually be tracked down in the event that it needs to be. This feature helps for information to be disclosed when necessary or urgent.

The concept of erasing messages can sometimes cause tension among other individuals, particularly when it comes to relationships. Many people view this concept as sneaky. Whether or not a person's intentions are good, it is important to respect another person's privacy. By doing so, tension will decrease, and individual privacy will be secured.

Deleting messages can come in handy especially when dealing with celebrities, like Paris Hilton and Tiger Woods. There is most likely information they share through text messaging that they wouldn't want others to get a hold on. The concept of deleting messages helps respect their desire to keep things to themselves, secure sensitive information, and keep the public from getting into their business.

It is important to remember that when you press the delete button in attempt to erase a message, the message isn't gone forever. Most cell phone carriers keep data in their systems for a certain amount of time before they are overwritten. It is safe to say that a person must be careful what they say through a text message, because they can be recovered. For instance, if someone was accused of being involved in illegal matters, investigators can trace back text messages sent to verify information. This has been done numerous times in order to verify the validity of a situation. Although some may view this as an invasion of privacy, it is a feature that helps catch people who are doing things wrong.

Whatever the reason, deleting messages enables people to choose what information others can get a hold of on their cell-phone. There are several reasons people give for doing so; it is a good feature to have. Erasing messages is an option people have, not required or essential. This concept is all up to personal preference, but mainly respects and protects the privacy of each individual.

Emoticons

The word "emoticons: comes from the two words emotion and icon. Therefore, the meaning of the word emoticon is an icon that shows emotion. Emoticons are often used in texting to show the sender's emotion. The emotions portrayed through the emoticon can be anything from happy to sad or even surprised. Emoticons are combination of text keys that put together form a sideways face for example :), a happy face. There are also a couple of special emoticons that are not sideways faces (Ex: <3, a heart). It is appropriate to use emoticons whenever you are trying to show your emotion through your text message, or you want to set the mood of a text conversation. For example if you want to show that you are just kidding around with someone, putting a smiley face, :), at the end of a text would show that.

Example Text Messages
with Emoticons Used

Texter 1: So ill cya at the mall tomorrow at 2?

Texter 2: Yep cya then :)

Texter 1: Guess what happened to me today :(?

Texter 2: What?

Texter 1: I lost my wallet at the mall >:(

Texter 2: Oh noooo :o

Texter 1: I ttyl

Texter 2: bye, I <3 u!

Texter 1: <3 u too!

Emoticons, How To Make Them

And What Each Means

Emoticon	Buttons used	What it means?
:)	Colon, closed Parentheses	Happy
=]	Equal sign, closed bracket	Happy
:(	Colon, opened parentheses	Sad
=[	Equal sign, closed bracket	Sad
:o	Colon, capital O	Surprised
:P	Colon, capital P	Silliness or frustration
:\	Colon, slash	Worried
:x	Colon, capital X	Not saying anything
:D	Colon, upper case D	Excited, very happy
:l	Colon, lower case l	No Comment, indifferent
:,(	Colon, comma, opened parentheses	Sad and crying
;)	Semi colon, closed parentheses	Happy with a winking eye
B)	Upper case B, closed parentheses	Happy with sunglasses on
>:(	Right arrow, colon, opened parentheses	Angry/Mad
<3	Left arrow, number 3	Heart
</3	Left arrow, slash, number 3	Broken heart
-_-	Dash, underscore, dash	Annoyed

Glossary Of Texting Abbreviations

And Their Meanings

Abbreviation	What it stands for	Abbreviation	What it stands for
A.S.A.P	As soon as possible	imu	I miss you
A&F	Always and forever	jc	Just chilling
bamf	Bad a** mother f***er	jk	Just kidding
bf	Boyfriend	jp	Just playing
BYOB	Bring your own beer/bottle	jw	Just wondering
dl	Down low	kit	Keep in touch
dtd	Down to drink	lmao	Laughing my a** off
fb	Facebook	lol	Laugh out loud
fu	F*** you	nbd	No big deal
FYI	For your information	nmu	Nothing much, you?
gf	Girlfriend	np	No problem
gn	Goodnight	omg	Oh my god
gtl	Gym tanning laundry	rofl	Rolling on the floor laughing
g2g	Got to go	smh	Shaking my head
hbu	How about you	stfu	Shut the f*** up
hmu	Hit me up	ttyl	Talk to you later
idc	I don't care	ty	Thank you
idk	I don't know	wbu	What about you
ilu/ily	I love you	we	Whatever
ilu2/ily2	I love you too	wtf	What the f***

Chapter 3

Facebook Etiquette

What is Facebook?

Facebook is a social networking site that most teenagers and now adults depend on to stay in touch with friends and family. Facebook was founded in 2004 with the original name "The Facebook" on the campus of Harvard University. To learn more about how it started watch The Social Network (2010) produced by Columbia pictures and written by Aaron Sorkin. The movie is loosely based on the true story on the development and unexpected success of Facebook· It has everything from pictures, to statuses, to games and more. Everyone who creates a free account has a "wall" where people can talk to you, post videos, and links to other websites. For most teenagers and young adults, Facebook is their connection to their friends, especially in college. Many people also use Facebook to get jobs and create business through the connections. Facebook's importance varies from person to person. Mostly it is a connection and an easier way to socialize with people. Facebook creators enjoy changing the layout every few months, so be aware that it will happen.

Privacy

Facebook is an amazing internet site. It can connect anyone throughout the world. Even though it can connect people from all over the world, everyone wants their privacy and parents are concerned about safety. It can be uncomfortable for a person that's unknown or not a friend to look through another's Facebook. That's why there is a privacy setting for one's Facebook page. This privacy setting lets you decide what other Facebook users can see on your page. When creating a Facebook page, Facebook asks what other people are allowed to see. There are many different topics that one can control who can view them. The topics include status, photos, posts, bio and favorite quotations, family and relationships, religious and political views, birthday, comments on photos, and contact information. The Facebook user can select who can view all of these topics. They can choose from only friends, friends of friends, or everyone. Every topic is individual so one can choose who sees what topic. The Facebook user can create his/her own privacy setting and change that at any time. Facebook has recommendations to help you suggest privacy settings that are good for you.

With a site like this that can connect a person to everyone around the world, there will be people that might follow a specific person's Facebook too much. The people that follow and always look at one particular person's Facebook are called "creepers." These creepers will follow all the Facebook activity for one person. They will look at their profile page all the time, always look at their photos, and will always comment on either their statuses or photos. If someone feels they have a creeper, there are ways to block him/her. One can delete them from their friends on Facebook and change their privacy settings so that only friends can see and comment on their Facebook page. One can also just block that certain person so that they cannot see any of their information.

Friends in common with another Facebook user are called mutual friends. One can look at the mutual friends you share with another person by going to the other person's homepage and there will be a mutual friends list. Mutual friends are important for friend requests. If one receives a friend request and they aren't really sure who the other person is, they will look at the friends that both people share. If there are many mutual friends and that person is comfortable with it, the friend request will most likely be accepted.

Status Updates

A very popular feature on Facebook is the "status." Everyone's profile and homepage consists of a status. When someone wants to "update your status," they can make their status consisting of anything they want. A person's friends can see all the statuses that are updated. You can also "tag" people in your status;

for instance, one can write in their status, "Going out with @ MarySmith tonight, can't wait!" When someone creates a status with the "@" sign followed by a name, it gives a link to that other person's page and includes them in the status. A status can be dealing with emotions, physical things that must be done, lyrics from a song or something else significant to them, things that happened that day, and unfortunately negative events that happen in someone's life.

There are different kinds of statuses. Statuses can consist of absolutely anything the person wants to post. There are "happy" statuses, where you clearly display your emotions through happiness and there is nothing negative about your status. Depressing statuses seem to be the focal point of statuses. If someone is depressed or does not like something about their life, they will post it up on their status. For instance, if someone was upset about a grade they received, a status would look like this; "Ugh I'm so upset about the grade I got in Psychology. Couldn't be more upset." Basically, any kind of emotion can be the focal point of a status.

People usually update their status with what they are doing. It could either be something they did in the past, something about the future, what happened that day or in previous days, or simply what they are currently doing. For example, someone's status can say, "Class until 2:15 PM and then dinner at 5:00 PM. Text me later!" Along with the status, they can also say that people can or cannot contact them. A downside to this is the fact that some people may find it annoying when it comes to friends who update their statuses every 5 minutes about what they are doing or how they are feeling, and it gets to be quite annoying to others. People sometimes write a paragraph on statuses, and yes you can post whatever you'd like, but it gets quite redundant and annoying when people write a whole paragraph on whatever it is they're talking about. People usually don't care or need every detail on everything. People can also be dramatic when it comes to updating their status. Some people may actually use statuses to get attention from others. The meaning of the status can either actually be towards someone to get their attention, or merely just have their status be anonymous and not focused towards anyone

51

in particular. People also tend to be very dramatic when it comes to updating statuses, causing a lot of drama between either a friend or a boyfriend/girlfriend. Suffice to say that status update themselves can create their own conversations and issues for the author and their friends.

People may also update their statuses on how they are feeling or how they felt about something. It can be about who or what they are currently mad at, and these particular statuses are dealt with emotion. Along with feelings and emotions, someone can also update their status with lyrics that are appealing to them. That is a very popular aspect of statuses. Song lyrics on someone's status can simply be about how they're feeling, it can be about an ex-girlfriend or ex-boyfriend, a current girlfriend or boyfriend, about family members, a situation someone is in, about their life, or simply just because they enjoy the lyrics.

A very popular status update would be when people display their emotion and respect in regards to recent deaths. This would include the deaths of family members, friends, and for some, famous people as well. They do this by writing, "R.I.P" followed by a date and a quote or saying that is meaningful to the person. When friends of the person displaying this status see this, they have the option of "liking" the comment or commenting on it. Usually, that can show a sign of respect.

One major negative of statuses are depressing statuses. When people are depressed and want to seek attention, they will display their depressed emotions through their status. These statuses are made to be dramatic and catch the eyes of others, to possibly spark a reaction out of the attention seeker. These statuses are very upsetting, scary, and sometimes lead to suicide threats. The only thing that comes positively out of suicide/depressing statuses is the fact that people can report it to the police and possibly help them, or the police themselves can see it. Sometimes it results in a suicide occurring because friends didn't read it or were not fast enough.

Pictures

Facebook is different from many other social networking sites because of the use of pictures. The uploading, tagging, commenting, and content of the pictures are a big part of this social networking site. Not only do you have a profile pictures but you can upload albums which can be pictures of one specific night or event. You can also tag people in pictures that they are in from your album or picture upload. Facebook allows you to comment on your friends pictures and also allows the feature of liking a picture by clicking the "like" button. This is another way of communicating with others on Facebook. People are able to have full conversations right underneath their picture within picture commenting. Pictures on Facebook are a fun interactive way of posting information in non-verbal ways and getting feedback in a semi-safe environment without the fear of being exposed.

Profile Pictures

Everyone with a Facebook account needs a profile picture to display when you navigate to their home page. The content of your profile picture is up to you. Some people choose to have a picture of them individually while others tend to have group pictures where it's a group shot of you with all your friends. Other times, people don't have picture of themselves and choose to post pictures of favorite sports teams, bands, or cartoon characters as their profile picture, usually to raise awareness for some kind of cause. Any one of these choices is acceptable in the Facebook world.

A profile picture is actually a pretty big deal on Facebook. It is the first thing people see when they are browsing on your mutual friend's pages. In this day and age, the internet plays a big role in meeting people. Unfortunately, this isn't the absolute safest way to meet people but it does happen quite often. Facebook is important in meeting new people. So think about it, your profile picture is the first thing they will be seeing. People tend to post pictures that they feel they look good and where they feel confident. Other people tend to put up a picture showing off their sense of humor displaying an awkward or funny face. Either way, the profile picture is the first thing people notice about you and unfortunately have a tendency to judge you on. So be careful what you post! First impressions tend to last. Many people friend other students at the colleges they are accepted to and may talk all summer with their roommate before they meet. All they have to look at is the profile pictures and albums, making an impression and hopefully attracting new friends.

Profile pictures also attract people to comment and like the picture. You will see that some people's profile pictures get more attraction then others. The picture usually gets the most comments and likes when it is first posted. On your homepage, it is easy to identify when your friends have changed their profile pictures. It will show a bunch of friends bunched together displaying the new picture in a small thumbnail with a link to a bigger version of it. Depending on the person's circle of friends or their personality is what will result in the number of comments and likes their picture will receive. Generally girls tend to get more comments on pictures and likes from their best friends or guys that may be interested in them. Facebook is a social network that is used to meet people that could, in some cases, become more than friends. Your profile picture could be the first impression you have on a person, so think about the content wisely.

Pictures, Albums, What and

What Not to Post

Your profile picture on Facebook is one of, if not, the most important thing. It is the first thing that people see when they surf to your page; you must choose your picture very carefully. Of course, you want to choose a picture that you look the best in, but let's not be narcissistic here, don't edit yourself to look like a Barbie! You have to remember that mostly everyone who is viewing your page knows what you look like so it will be a bigger embarrassment when your picture looks nothing like you than you think. Another rule of thumb when it comes to profile pictures is to remember that just because you look good in the picture does not mean that your friends do too and therefore you should rethink using that picture. Just think, you would not want an embarrassing picture of you floating around the internet! There are many different types of "good" profile pictures; it can be a posed picture or one that someone took of you when you were not even looking. One of the worst things you can do is change you profile picture multiple times a day, right there is a sign that you have too much time on your hands and you should find something else to do with all of your spare time.

Even though your profile picture is the most important picture on your Facebook wall does not mean that you should neglect all of your other pictures. You still need to worry about the pictures that people are tagging you in and the albums that you are creating. You are the judge on what to post publically and what not to post. A personal rule of thumb to follow is to never post something you would not want your grandmother to see. Not only do you need to respect yourself and not post incriminating pictures but you still have to respect the other people in the pictures. You will take plenty of amazing pictures in your life, not posting one good one of you because it is embarrassing to others will not kill you.

One saving grace that we all have on Facebook is the "remove tag" button. If you are ever tagged in a photo that you don't want people to see when they go onto your wall, all you have to do is simply untag yourself. This removes your name from the picture and it will not show up in any of your albums!

Poking

Poking is a feature on Facebook that is on your profile, in which someone can click on it if they want to catch you interest. When you receive a poke, it comes up on the home page when you log on. When someone receives the poke, they have a choice to either poke back or to remove the poke. When someone does not receive a poke back, it is like a slap in the face because they do not care enough to poke back or they do not want to have that connection with you. The act of poking can be perceived in various ways because there is no direct meaning to it. It depends who pokes you, and it is up to you to figure out what the meaning is behind it.

The following can be the reasons as to why someone poked you:

☐ Friends do it to be annoying or just for fun

☐ Friends often partake in a poke war, which is to see how long they can poke someone before they get sick of it and want to stop

☐ A way to be flirty

☐ An excuse to initiate conversation between people

☐ A way to say hi or what's up without actually saying it

☐ It can be a start to a relationship because it leads to people talking (like the boy and the girl in the playground during school that hit each other because they like each other)

☐ Be aware: it is creepy if someone pokes you that you do not know or they are not even one of your Facebook friends.

So, when you receive a poke from someone, you can use these reasons to figure out what they intended. The best way to find out the reason is to poke back and ask!

Facebook Relationship Statuses

Did you ever request to be someone's friend just to find out if they were taken or not? One of the major reasons Facebook is such a success is due to relationship statuses. People will friend request others just to see what their status is. Today in our generation, to be fully in a relationship, people don't believe it's real unless its "Facebook official". Relationships can be seen so quick now due to mini feeds and how often people check their news feeds. News feeds are on the main page of Facebook that shows all new activity done by your friends. The second a relationship is made

"FBO" (Facebook official) it is open and public for all of Facebook to see. Having it so public though can be stressful. When a relationship is over, people can know this right when it ends. This can cause either support by people, or start gossip.

There are a few things that are clearly not cool. First is changing your relationship status on Facebook before telling the person you are breaking up with them. Nothing could be worse than finding your relationship changed via the status change. Changing your relationship status just to mess with the other person is also not cool. Both of these items create more drama than they are worth and really have no purpose other than to draw attention to you.

Facebook relationship statuses are becoming more and more in depth recently. There are now more options on what you want your relationship status to be. Options are single, in a relationship, engaged, married, it's complicated, in an open relationship, widowed, separated, or divorce. There is clearly an option for every type of relationship you could possibly be in. But don't always believe every status you see.

For most new "Facebookers", they find humor in making fake relationships. Often you will see a girl and her friend married or in a complicated relationship, when they are really not. For the newer relationship, they will never just leave it single; it will most likely be a fake relationship. You will often see couples put that they are married or engaged. It is often funny, but can be taken seriously and could be a problem. For the older generations on Facebook they have the proper relationship status.

For relationships today, you know it's serious when it's FBO. This goes to show how much of an impact Facebook has made in relationships and how something so simple like changing your single, to in a relationship, to possibly some other status something else can have such an impact on your love life.

Facebook Quiz 1

Not sure whether or not to make your relationship Facebook Official? Take this short quiz to see the answer!

1. Has the boy/girl officially asked you out? (Yes or No)

2. Do you and your significant other tell people about your relationship? (Yes or No)

3. Does the relationship seem somewhat serious? (Yes or No)

4. Do you and your significant other seem to show PDA? (Yes or No)

5. Do you and this person have pictures that show you're in a relationship or do you write on each other's wall in a loving way? (Yes or No)

...If you have yes for these answers then congratulations, it's time for you to make it official!

Facebook Do's

1. Use it for social networking... that's why it was created.

2. Post pictures right away... people get annoyed if you don't.

3. Tag your friends in pictures... it saves them the trouble and is easy.

4. Friend classmates... just in case you need to know an assignment.

5. Block creepers... don't add them either.

6. Look at mutual friends... you may have mutual friends but those friends may not be "real" friends.

7. Utilize invites... it's easier than mailing them.

8. Change relationship status... it has to be Facebook official.

9. Block ex's... don't drive yourself crazy stalking them or having them stalk you.

10. Be open to meeting new people but be careful around them... friending strangers can be dangerous.

11. Update your status about good news... that is things you want to share with people.

12. Stay away from people you were once involved with on your page.

Facebook Don'ts

1. Don't Bully, pick on, or make any character assassinations of people you are enemies with or don't like. Consequences for you and the other person can be lifelong (e.g. emotionally scar, the person kills themselves, comes after you and kills you, or you get arrested and have a permanent record).

2. Put pictures up of drinking.

3. Post things you wouldn't want your parents or grandparents to see... they are on Facebook now too.

4. Stalk people... you will drive yourself crazy.

5. Post creepy things on peoples walls... they will "de-friend" you.

6. Change your status every two minutes... they will block you.

7. Un-friend friends during a fight... you will regret it later.

8. Post things that should be private... people do not need to know everything.

9. Friend strangers... it's like talking to a stranger.

10. Send requests to people who don't use that application (like Farmville).

11. Use Facebook to make excuses.

12. Threaten people on Facebook... it's called hard evidence.

13. Make status about things people do not want to know about. Example: I need to pee.

14. Friend request your boss… because you know you will complain about your job on it.

15. Use it for dating. That's what dinner and a movie is designed to do.

16. Post things you will regret later… trust me you will.

17. Talk about people on other people's wall… it will get back to them.

Quiz 2: What do others think of your Facebook use?

1. How often are you on Facebook?

    A. Once a day

    B. A few times a day

    C. Once a week

    D. I don't go on

2. Do you poke?

    A. Only my significant other or best friend

    B. I poke everyone

    C. I don't poke at all

    D. I poke a few people

3. How fast do you upload pictures?

    A. As soon as I get home and I tag everyone

    B. The next day and I tag a few people

    C. A week later and I only tag myself

    D. I don't upload pictures

4. How often do you update your status?

    A. Every five minutes

    B. When something good happens

    C. A few times a day

    D. Once a month

5. What do you post on your statuses?

    A. I don't post anything

    B. Song lyrics

    C. Everything I do

    D. Good news

6. What do you post on someone's wall?

    A. Personal information

    B. Just saying hi

    C. I don't post anything

    D. Links to songs or videos they would like

7. Who do you accept friend requests from?

    A. No one

    B. Strangers

    C. Only people I know

    D. People who I have mutual friends with

8. What is your privacy setting?

    A. Only my friends can see my profile

    B. No one can see my profile

    C. Everyone can see my profile

    D. I block certain people from seeing certain things like my parents or an ex

9. What does your relationship status say?

    A. It says I'm married to a friend when I am not

    B. It says whether or not I'm in a relationship

    C. I don't have it showing

    D. I change it every day

10. What do you use Facebook for?

    A. Social networking and staying in touch with friends

    B. Stalking people

    C. I just have one because everyone else does

    D. To be social

Points

1. A= 3 B = 4 C = 1 D = 2

2. A= 4 B = 2 C = 1 D = 3

3. A= 2 B = 4 C = 3 D = 1

4. A= 2 B = 4 C = 3 D = 1

5. A= 1 B = 3 C = 2 D = 4

6. A= 2 B = 4 C = 1 D = 3

7. A= 1 B = 2 C = 4 D = 3

8. A= 3 B = 1 C = 2 D = 4

9. A= 2 B = 4 C = 1 D = 2

10. A= 4 B = 2 C = 1 D = 3

Score between 30 and 40

You are everyone's favorite person on Facebook. You do everything correctly. You are not excessive with Facebook but you know how to use it without people getting annoyed at you. Facebook is very natural to you to use correctly.

Score between 20 and 30

You sometimes have to work at using Facebook correctly. You try to fit in and want to connect with people with Facebook. Your friends enjoy how much you stay in touch with them on Facebook.

Score between 15 and 20

You post too much on Facebook most of the time. You need to

learn to be more secretive with your life on Facebook. Try taking a step back Facebook.

Score between 10 and 15

You do not go on Facebook at all. Try going on Facebook a little more and being more social on it. Take advantage of all things Facebook provides to keep you connect to friends and family.

*Disclaimer: The quizzes in this book were designed as fun and are not to be considered as valid or reliable.

**Important notice: Facebook is prone to changing and will change without notice by the creator!

Conclusions

In this book, we hoped to provide information to the new social networking users, some guidance while reminding you of the "pros" and "cons" of social networking, and some reminders and issues that should be thought about. It is our desire that social networking users found this book enjoyable and learned something in the process. As technology continues to change and our use of cell phones, texting and Facebook evolves, please remember the goal of social networking is to connect with the people who are important to us and technology is just a means to an end. The relationship is what is important. We look forward to seeing how networking changes and improves. We are really glad that you took the time to read *The TXT book: the guide to social networking*.

Terms

Digital Native- those that were born into the age of technology, a cell phone on their ear and their hand on the key board.

Digital Immigrant- those born prior to the idea of cell phones, texting or a computer that checked your grammar and spelling. The learning curve was intense.

Facebookers-a word we might have created to mean those that use facebook.

Texter- another word we think we created to denote those who text.

The MU Socially Connected:
(The authors)

Top Left to Bottom Right:

Danielle Long, Dillion Fields, Amanda Kruzynski, Emily
Prinsell

Jill MacConnell, Axhi Popinara, Kelly Rose Printon, Kirsten
Kearns

Gabriella Tornatore, Mary Fulco, Rachel Conners, Kaite
Ellershaw

Alyssa Stevenson, Mary Dates, Lisa Sypniewski, Erin Smith,
Demi Moore

Sara Pietrowitz, Lizzi Rowe, Ava Pignatello, Danielle Medina,
Kelly Ward, Ricky Hemberger

Missing: Malory Jasmin and Dan Wentel